RIVERS
OF THE
WORLD

Lifelines of Mankind

I0504121

DR.THIRUMENI

INDIA · SINGAPORE · MALAYSIA

Notion Press Media Pvt Ltd

No. 50, Chettiyar Agaram Main Road,
Vanagaram, Chennai, Tamil Nadu – 600 095

First Published by Notion Press 2021
Copyright © Dr. Thirumeni 2021
All Rights Reserved.

ISBN 978-1-68523-442-3

CONTENTS

LIFE LINES OF MANKIND OR RIVERS OF THE WORLD PROLOGUE

Five elements are essential for human life and civilization. They are land, water, air, sky and fire. Without land man can not produce crops and fruits. Without water he can not irrigate his lands. Without air he can not breathe. Without fire he cannot cook his food and prepare his hot drinks. Without sky or space he can not receive light and heat of the sun or glow of the moon and stars.

These are the gifts of Nature or of God to man. In that case to preserve these things and to protect them for sustainable existence is the first and foremost duty of mankind. Animals kill for hunger. But man kills for sport and pleasure. Hence today man is the predator. Man does not live alone on earth, nay he can not live alone on earth. He depends on other creatures for his life. There are other species. Animals are there, birds are there and plants are there. They are very helpful to man. Cows give us milk, horses draw our carts, bulls are necessary to plough our lands, plants give us fruits and flowers.

But man, shrewd man exploits nature to full extent and destroys it totally without any idea of future generations. Man is the prime animal on the earth, but he should share it with the birds of the air and the beasts of the woods.

The Theory of ecological balance is advanced by me with the vision of continued human life on this planet. No other planet is good or suitable for human life or habitation. Hence man should preserve this planet with care and love. Only with this theory of ecological balance and that principle of environmental protection I have written this book on the rivers of the world.

These great rivers are life lines of man and society and they are the foundations for various civilizations.

It may be Egyptian or European. It may be Mayan or Aryan. All the civilizations thrive on the banks of the rivers. Rivers are my best friends and my life has been spent on the banks of rivers like

Vaigai of Madura, Cauvery of Tanjore, Kedilum of Cuddalore and Adyar of Madras.

In the following chapters main rivers are analyzed from various angles. Their origins, courses and mergers in the seas are described in details. Water is everywhere. It may be wild and violent Atlantic Ocean. It may be gentle and quiet Pacific Ocean. All waters of the oceans are the same.

CHAPTER 1

Our earth as a planet moves in space, but it floats on water. Hence the primacy of water in human life is emphasized and the uses of rivers, their pollutants and their contributions to trade and transport are explained in the following pages.

Fish is the child of water and man is also the child of water and death of man by water, for water and in water is mentioned in all the scriptures of the world.

Rise of Civilizations:

The world is the gift of God to man. The human species exist only on earth and not on any other planet or stars. The existence of aliens on other stellar regions and their visit to our earth in strange vehicles is only a conjecture and not a proven one. Our space adventures on Moon,

Mars, Saturn and Jupiter may be successful but not useful. In those celestial bodies we don't have enough water or air or suitable temperature. The American astronauts when they returned from moon remarked that the earth is the most beautiful object in the universe.

The world contains the elements which are essential for human life and civilization but they are not under human control. Even other species like animals, birds and reptiles and aquatic creatures exist only with the help of these essential elements. Fish dies out of water, man dies when he loses breathing, trees die without water. Even ferocious animals like lions, tigers and leopards can not live without hunting the lower animals and they too wander for water in hot seasons. The basic theory is that these five elements are the greatest gifts of heaven to man and when man spoils them he digs his own grave.

A polluted earth, spoiled atmosphere and poisoned water will ultimately spell the doom of mankind and animal species. Hence man, to protect his species and to prolong his existence is bound to know more about earth, water, air, sky and fire.

If these things are destroyed man's life is ended, the Biblical forecast of deluge and final judgement will prevail. Among these five elements man needs earth to walk on, air to breathe in and water to drink. Without these elements man's life is rounded with a sleep. Hence to underline the pre-eminence of water and water resources this book is written with a special emphasis on environmental protection.

Rivers are living things, they have their life and movements. Some of them have faced their death in the past.

Rivers are not water lines. They are our life lines. Rivers are the cradles of civilizations and creators of various empires. Man could sustain himself with comfort and commodious living only on the river bank.

We can not forget Egyptian civilization on the Nile, Mesopotamian on the Tigris, the Dravidian on the Indus, the Aryan on the Ganges, the Chinese on the Yangtze and the Mayan on the Amazon. Kingdoms and empires were built on the river banks, temples were erected on these riversides and festivals were celebrated close to these rivers.

As blood is to the body, water is for human civilization. Oceans are store houses of water that is salty. But the vapors rise from oceans, become clouds and pour down as rains on mountains and lakes. The rivers emerge from these mountains and lakes and flow on the plains giving life and stability to human civilization.

Hence the best and the biggest rivers are chosen across the five continents and their origins, movements and ends are described in details. This book is a guide to rivers, their courses, their uses and their mergers. Some rivers are wild, a few rivers are gentle and very few rivers are dangerous.

But rivers are rivers and without them we are nomads, hunters and barbarians. They give stability to human settlements. In India rivers are symbols of femininity. Some rivers are holy and divine. Hence Indian approach to rivers is awesome and attitude is reverential.

Illustrations and Images:

1. Mountains and peaks with snowlines.

2. Rivers meeting oceans and waves.

3. Towns and cities on river banks.

4. Swiss Alps – Andes – Rockies – Himalayas.

5. Danube and Black Sea – Amazon and Atlantic.

6. Nile and Mediterranean Sea – Zambezi and Pacific.

7. Madurai on Vaigai, Madras on Adyar, Rome on Tiber, Budapest on Danube St. Louis on Mississippi.

CHAPTER 2

EUROPEAN RIVERS

Danube River:

As an Indian or a Dravidian, I should choose Ganges or Indus for my first chapter. But I have chosen Danube of Europe as a first one. This is because this river has enchanted me for a long time with its blue colors and musical flow. As a matter of fact, I have chosen three rivers from each continent totally covering fifteen rivers. However, for Europe and Asia, I have chosen four rivers each and thus increasing the total from fifteen to seventeen.

During my European travels this Danube enchanted me in various ways and I saw it in Vienna, Bratislava and Budapest. Indeed, the Danube divides the Hungarian capital into Buda and Pest. Danube starts in Black Forest and ends in Black Sea. The travel from Black Forest to Black Sea is very long and slow but very pleasant.

Danube creates modern European civilization with its ships transport, boat rides and cruise tours. It covers more than ten countries and has many tributaries. Germany, Austria, Slovakia, Hungary, Serbia and Romania are a few countries to mention.

The story of Danube is the story of European art, culture and music. It unites Europe, it divides countries and it embellishes the European landscapes. The Danube has many names according to the countries it passes through. It is famous and popular in Vienna because of music programs in that city. Many composers have made music in the name of this river. Great musicians like Mozart and Beethoven are products of this river side. Even the dances in opera houses of Vienna and Salzburg are the echoes of the music of this wonderful river.

It travels more than three thousand miles. It goes through small countries like Ukraine and Maldova. Danube basin is the agri-basket of Europe. In the end portions Danube creates many marshy areas which are breeding grounds for various birds. Many kinds of migratory birds stay here for a few months and the delta is noted for biodiversity. It was the eastern border of Roman Empire in the past.

Apart from agriculture Danube is very helpful for ship movement and tourism. Danube river cruise is very famous among the Europeans and one can see all major cities and countries of the continent on the way.

It is called the Amazon of Europe for its long journey. Danube enriches European culture and civilization in many ways. Wonderful cities like Vienna, Belgrade and Budapest can be seen on the banks. Once it was a place of ancient human settlement. Indeed, European first settlement originated in Danube areas. In the medieval times the river was famous for fishing. Now due to various needs of men it has taken many shapes. It has many hydro electric projects. Gate of Iron is an important stage in the course of the river. It has been declared as a heritage site of UNESCO. One cannot see Europe without Danube. Europe without Danube will be like a desert. What is Ganges to India Danube is to Europe.

Danube is the home of fresh water fish system. On the way river creates forests lakes and islands. This is the most powerful river of Europe. It moves under the ground in Black Forest for thirty-five kilometers. Indeed, there is war between Danube and Rhine in the starting points. While Rhine moves to the North Sea, Danube moves to the Black Sea.

Many steps are taken to protect this river from pollution. Eastern Orthodox Church archbishop and others have organized groups to reduce pollution to retain its natural course. In Budapest and in other places there were protests to stop diversion of this river course and creation of small canals. Environmentalists, social activists and Christian priests have come together to show the value of this river to the common people and to governments which try to exploit it for energy and oil. A man or leader must walk along its banks to maintain his popularity among the common people and to unite the countries of the continent.

Volga River:

Next river of name and fame is Volga which passes through Russia. Volga is the longest river of Europe. It is a special river for Russia. It is Russia's national river and its beauty and benefit are described in Russian literature and folklore.

Volga passes though many important cities like Karson, Moscow and Volgograd. It is rich in fish culture and trade. It has many tributaries which are more than two hundred in number. Volga empties in Caspian Sea after travelling more than two thousand miles. Its delta is very vast and wheat and other crops are grown here. Volga has many dams and dykes which help in irrigation of crops and production of electricity. Volga moves very slowly along the plains and ship movement and cruises are possible. It is highly polluted by industries located along the banks. It drains the western part of Russia and as Indus is to Pakistan Volga is to Russia.

This is the life line of Russia and one cannot imagine Russia without Volga as one can not imagine Rome without Vatican. Many species of birds, animals and aquatic creatures can be found in the estuary of Volga. The word Volga in Slavic language means water. It has a suspension bridge also. It is so deep and wide that we cannot see the other side on certain locations. It has three divisions upper, middle, and lower.

It freezes for three months in winter. Sturgeon is the main fish. Volga grad was the battlefield in the second world war. It flows from melting of ice and its dams control floods. But today Volga has become dry in certain areas, it is like a desert. The hydel projects have stopped flow of water and many ships and boats have to wait to cross bridges.

Bolsheviks in their mood of euphoria built so many dams and stopped flow of water. Not only that, they started industries for the workers and these industries polluted the rivers totally and killed all the fish culture. Only when rains fall, water level in the river is up and the ships and the boats move. This is the tragedy of the National River of Russia. Communist rule has destroyed the river too much. Not only that these communists have spoiled wonderful churches and beautiful monasteries in the cities of Russia.

Rhine River:

Rhine and Rhone rivers are wonderful and beautiful. Rhine is a pride of Germany while Rhone is a pride of France. Rhine river cruise is famous in Europe. It is called Rhine Romantic cruise. In the medieval period Rhine banks were occupied by robbers. They collected tolls from the ships of traders. So, we can find castles on both sides of Rhine River where these robber barons thrived.

Rhine falls near Swiss town of Zurich are very famous. The falls are romantic and there is a castle on the other side. We could see the Rhine River area filled with boats of tourists who are adventurous to go near the falls. This river goes to North and merges in the North Sea in Netherlands. Rhine River passes through many countries like Austria, Germany and Netherlands. It starts in the Alps of Switzerland. Famous towns are located on its route. The towns like Cologne and Zurich are close to this river.

If Danube is European in the eastern side, Rhine is European on the Northern side. Rhine is also polluted like other rivers. It is due to location of chemical factories on the banks. It took many months to reduce this chemical pollution. The river was full of fresh water and Salmon fish was in abundance. But now due to pollution the Salmon population is highly reduced.

Rhine cruise is described romantic because of castles on the banks. Rhine is a traders' route carrying a lot of coal to and a lot of oil from North Sea ports.

Generally European rivers behave in a common way. They flow gently and slowly and there are no rapids in their passage. But African rivers on the other hand move fast and the rapids are hindrances to ship movements and boat rides. They are good for ship transport but sometimes locks are located to balance the water levels. East flowing rivers go to the Black Sea. South flowing rivers go to the Mediterranean Sea. West flowing rivers go to the Atlantic Ocean. North flowing rivers go into the North Sea or the Baltic Sea.

Rhine River is described as the graveyard of ships because during Second World War ships were sunk in the river to avoid bombing by the enemies.

St. Goar is a beautiful town on its banks, and one can find crowds of tourists in this town because of this river cruise. One beautiful city is noted for EU parliament. That is called Strasburg. Another city is Cologne which is noted for high cathedral. Wherever Rhine flows, there flows the tourist group.

The Rhone River:

This river is essentially French. This cuts across France and flows into the Mediterranean Sea. Rhone moves slowly and the tourists enjoy on both sides vineyards and buildings of medieval period. Roman settlements are famous in these areas. Roman arena Roman aqueducts are here on the sides of the Rhone River. Rhone valley is famous for three things. The best vineyards can be seen in Burgundy, the wine country of France. The best types of wine are prepared by the locals and tasted by the guests and the tourists. Rhone flows in the south east of France along the banks of which there are walk paths and cycle tracks.

On the banks of Rhone there are famous cities like Lyon and Avignon. In the medieval period there was difference of opinion over Papacy and French pope was living in Avignon for fifty years. The three famous things of Rhone valley are Pope's palace in Avignon, Roman Amphitheatre and vineyards.

A man must have a Rhone River cruise to understand France and appreciate the French culture. Rhone starts in the Swiss Alps and moves towards South. It totally has the colour and nature of France even though it passes through Geneva Lake in Switzerland. Yet the French culture and its splendor can be seen not in Paris, a city of light and night clubs but in the towers of Rhone valley. An adventurous sailor can share his boat with his friend and can sail from the Mediterranean Sea and can reach North Sea sailing along the Rhone River.

CHAPTER 3

ASIAN RIVERS

Ganges or Ganga is the holiest river in India. Its origin is in the Himalayas and that place is called Gangotri. Gangotri is also a holy place and Hindus consider it as one of the four places along the Himalayas. The four places are Gangotri, Yamunotri, Kedarnath and Badrinath.

Ganges was a turbulent river of wild force and hence it descended from heaven with ferocity and fervor. It was Lord Shiva who controlled this river and it descended on earth through his locks from his head. Hence Lord Shiva is called Gangadhar, the lord adorned by the river. There is a story about it. The prince Bhagirath came to know that his ancestors were suffering in hell due to a curse. He had to redeem them and send them to heaven. Hence, he meditated for many years and brought the river to earth. His ancestors were washed by Ganga waters and they were released from hell and bondage. Hence Ganga is a holy river in India and once in twelve years the famous festival of Kumbmezha is celebrated on the banks of the river to show the spiritual significance of its water. Kumbmezha means pot festival – Kumb means pot and mezha means festival. This is celebrated on the banks of the river in a town called Haridwar a place where nectar was spilled by the celestials. Hence Haridwar is supposed to give immortality to man especially to Hindu. The festival goes on for more than forty days and beggars and saints, men and women thieves and sinners, masters and disciples the naked and the diseased assemble here and take a holy dip that washes off their evil deeds and sins and gives them divinity. The whole ritual is blessed by Lord Shiva who stands on the bank in the form of "big statue".

Ganga is a long river and passes through many states like Uttaranchal, Utter Pradesh, Bihar and West Bengal. Many holy places are located on the banks of this river. Some of them are Allahabad, Rishikesh, Benares and Haridwar. The hope is that dead bodies move to heaven to touch the God's feet. Hence it is a practice to throw dead bodies across the river in Benares, the holiest place for the Hindus. Hence along the banks of this river there are many Ghats or burning centers or cremation grounds for the dead bodies. You can throw the corpse into the river or you can bury it on the ground near the bank. The soul will reach the heaven and this is the faith of the Hindus. Another ritual is popular in Haridwar, on certain occasions and dates Brahminical priests perform certain rites to give peace to your dead ancestors. Thus, Hindus spiritualize the waters and rivers in all parts of India.

Ganges creates its own delta where wheat and paddy are cultivated. Ganga is a perennial river and water flows in summer and winter. In summer the ice on the Himalayas melts and water flows into the river. In winter floods due to rains flow in the river and many villages and towns are in peril.

Gangetic plains are the richest in India or in the whole world and wonderful crops and vegetables are grown on these fields. Famous empires like Magadha, Gupta and Vardaman were established on the banks of this river. It was possible due to its fertility and water. These Gangetic plains are the locations of Aryan civilization and Hindu scriptures were written here by the saints and sages. Last but not the least this ganga merges with Brahmaputra another from North East of India and empties in the Bay of Bengal. Ganga is holy but it is the worst polluted river due to tanneries and all the efforts of the governments at centre and state have become failures in the purification process. Hence Ganga is still a refuge of pollutants and garbage.

Ganga has wonderful tributaries on the upper reaches of Himalayas. They are Mandakini and Akilananda. But on the plains Jamuna and Saraswathi are tributaries. But the river Saraswathi is mythical and flows under the terrain. The confluence of these three rivers is in Allahabad and the meeting point is called Triveni Sangam where all rituals and rites are performed throughout the year by the pilgrims and the priests.

The holy river is sometimes dry in the upper reaches and many of the streams become sandy and one wonders at its pollution and draught. It is because of the environmental damage done to the Himalayas by the successive governments, tourists, pilgrims and mountain trekkers.

Indus River:

Our next Indian river is Indus of Dravidian civilization. It is called Indus civilization because of this river Indus. The name India is derived from the word Indus. Dravidians were the original inhabitants of India and they moved to north from south because of deluge and tectonic movement. Hence Indus civilization is the second stage of Dravidian settlement.

This river flows from a lake near Tibet of Himalayas and present-day Pakistan is its delta region. Because of partition in 1947 India lost many things and this river is a huge loss to India though it is a gain to Pakistan. Through this loss of Sind and Punjab provinces of fertility and wealth, India became suddenly bankrupt and beggarly. Indus has many tributaries and it is flooded in every monsoon season. Indus River merges in the Arabian Sea. It has major five tributaries. They are Jelam, Chenab, Beas, Ravi and Sutlej. These tributaries enrich various states of India. The states like Himachal Pradesh, Punjab and Kashmir get much benefit from these tributaries. For example Sutlej goes through Punjab. Jelam moves through Kashmir. Chenab and Ravi pass through Himachal Pradesh. These three Indian rivers Ganges Indus and Brahmaputra are gifts to North India.

They created fertile plains and rich empires. For example, the great Maurya Empire was the creation of the fertility of the Gangetic plains. If the Ganges is the basis of the Aryan civilization, Indus is the basis of the Dravidian civilization. They are opposite in culture and religious aspects. A simple example is enough: Aryans burn the dead but Dravidians bury under the ground.

Indus is the basis for the prosperity of landlords of Sind and Punjab provinces. The wheat is the rich grain from Indus delta. On the banks of Indus River, we can see wonderful cities of Pakistan. They are Karachi, Lahore, Mithakote and Attack. Even Peshawar is a noted city of this plain. The river starts in a Lake in Northern part of Tibet and travels more than three thousand kilometers. It has many dams and in monsoon it is flooded. In the year 2010 Indus floods destroyed major portions of Pakistan. However, without Indus Pakistan will suffer in energy and food production.

Mohanjetharo and Harappan civilization is based on the fertility of the soil of Indus River. Here there was mother worship. The symbols of Shiva and bull indicate the Dravidian settlement. Dravidian society is matriarchal and Aryan society is patriarchal. These proto-Dravidians were the original sons of the soil before the nomadic Aryans came from the Central Asia. They had trade relations with other regions of central Asia both on land and on this river routes. The Shivalik hills are nothing but sign of Shiva worship on this land. The Chandigarh Museum and place names in Pakistan testify the presence of Dravidians in the ancient times here.

Mekong River:

Among Asian rivers the best is the Mekong. It starts in Tibet passes through Laos, Cambodia and merges in South China Sea. Mekong delta is the rice bowl of Asia. Fish culture and trade are popular along Mekong delta. This river passes through six countries. They are Tibet, Laos, Burma, Thailand, Cambodia and Vietnam.

The river travels more than four thousand kilometers and mingles with South China Sea. This is the longest river in South Asia and its delta is rich in paddy cultivation and fish trade. Most of these countries are Buddhistic and they are proud of Mekong River. Mekong means mother of waters. But of late many friends of this river complain of two things. One is construction of too many dams. The other is location of industries near this river and the pollutants spoil the fresh water of this river.

Chinese have done maximum damage to this Mekong River. The Vietnamese and

Cambodians complain against Chinese dams and of less water flow in the tail end. Not only that because of pollution of the fresh water, the fish trade is affected and good fish are not caught in Cambodian and Vietnamese regions. The river touches three countries at one point at their borders. They are Laos, Burma and Thailand. People and lovers of earth are taking efforts to control and regulate the Chinese projects. They are amazed at the colour turn. Once the waters were red and muddy, but in certain areas waters are blue. This is due to dam projects and less water flow.

Generally, Asians are orthodox and superstitious. They hold rivers in reverence. But modern technological activities and hydroelectric projects have spoiled this wonderful river and it is painful to nature lovers.

A river cruise along Mekong is a delightful experience. It is adventurous and amazing. However, this life line of Asia is in danger and we have to restore it to its original shape, flow and colour. ASEAN is an Association for Asian development and unity. It has appointed a commission to save Mekong River. But nothing has been done so far, the commission has not taken any action on China. China is a big brother in Asia and this big brother treats younger and small brothers like slaves. China is a danger to Asia, nay it is a danger to the whole world itself. Countries like Thailand, Cambodia, Vietnam and Laos are having a dangerous brother in their neighborhood. Small countries like Bhutan, Nepal, Pakistan, Srilanka and Tibet will be swallowed by this big brother in course of time.

Mekong River is affected by sand-mining. The river sand is robbed for building and filling activities. Hence the river has lost its original shape and quality. The farmers and citizens of Cambodia complain about these things. There is a lake called Napien in Namphen, the capital of Cambodia. There are many lakes in and around this city. But building companies fill these lakes with sands taken from this Mekong River. Consequently, the people who live along the river banks and farmers are much affected and their livelihood is lost. But the government argues that sand is revenue and it gets lot of money by selling the sand to the construction companies. So, the river becomes empty of fishes and main losers are the farmers, fisherman and small boat traders. Big companies and multinationals destroy the resources of earth and governments are winking at their activities. More than that Singapore has extended its shore area by buying and bringing sand from Mekong River.

The poor people who cultivated small crops in these lake areas are much affected. They become refugees in their own land and the culprit is the government itself. In Vietnam the farmers are switching over to other crops because of lack of water. The whole Asia has been affected by the avaricious activities of Chinese Government.

Yangtze River:

This is called the mother of China and it is part of Chinese culture and civilization. It is a wild river running across China for more than thousand kilometers. This river has created a delta wherein Chinese produce much wheat rice and vegetables. The waters of this river floods are controlled by many dykes and dams. This river passes through many provinces of China. Its name varies according to the place it passes through. Very old settlements have been discovered on the banks of this river. It has many tributaries and consequently has very big delta. On its banks wonderful Chinese cities are located and one of them is Shanghai city. This delta is acute in grain production. Wheat barley and maize are produced in this delta. 70% of grain of China is possible in this delta. The river is dam-free and only very late a hydroelectric project has been completed. Ships can sail into the river from the sea up to thousand miles. It is very deep river and variety of fishes can be found in this river. It is also a very long river in Asia next to Mekong. It merges in East China Sea.

When China has got other rivers like Ho Yanko and Yellow, it should be satisfied with its prosperity and fertility. It need not disturb the rivers that flow from its borders to other countries. It is not a big fish to swallow the small fish. China is accused of spreading corona virus with evil intentions by American leaders. Under such circumstances it should avoid further accusations and charges from its neighboring countries of aggression and destruction of natural resources. It should follow the high philosophy of live and let live in future for its own benefit.

CHAPTER 4

AMERICAN RIVERS

Amazon is the most interesting river of the world. This is because of its length and its course though dense forest. Amazon forest is noted for rain fall biodiversity and tribal settlements.

The river starts somewhere in the Andes and moves towards east cutting across mountain ranges and high pleatues. Indeed, it is the rich river of Latin America. It passes through countries like Peru, Argentina and Brazil. Amazon is a very wild river and unfit for irrigation in high places.

When it comes to the plains, smooth river helps in transport, ship movement and fishing. The river merges in Atlantic Ocean. Its delta is not rich and colorful. Amazon fame rests on rain forests which bring not only rains but also stability to the South American climate.

It is called Amazon because it is an amazing river in many aspects. It is the river of rivers and wonder of wonders. Amazon rain forests are protectors of our earth and species of animals and birds. Amazon has more than one hundred tributaries. In certain places its width is more than fifty kilometers. Amazon covers 30% of earth as its basin. The waters of Amazon desalinate the Atlantic Ocean up to three hundred kilometers. Indeed, to day Amazon has become a point of dispute. On the banks of the Amazon the natural resources are exploited by the governments and private companies. Within two- or three-decades Amazon will lose its charm, if this exploitation is not stopped and those culprits are not punished in time. 20% of oxygen is produced by the Amazon forests. The volume of water emptied by the Amazon in the Atlantic Ocean forms 20% of the waters of the world. Indeed, Amazon is a single river which combines the Nile, the Mekong and the Mississippi in its flow of waters.

Amazon runs across the continent from the west to the east. Hence, we can call it a continental river. The city of Manus is located on this river banks. Indeed, Amazon is a mystery with its forests, tribes and animals. Amazon River is full of fish some of which are carnivorous. The biggest snake Anaconda lives in this river. Amazon has neither bridges nor reservoirs. Because it is big, it is called River Sea. Some of the tribes of Amazon forests never come into the social contact.

It is the longest river running from the west to the east covering the South American continent. It crosses various types of terrain. It crosses mountains. It moves on pleatues. It goes through forests. It moves over rocky region. Finally, it touches waves of the Atlantic Ocean.

Its journey is epic and grand and beyond the ken of men and women. It is a universe with animals and birds and it is the guardian of this world. A man or government which disturbs this Grand River system disturbs and destroys the world and the mankind in it.

Mississippi River:

Mississippi is a beautiful river of the United States of America. It passes through more than ten states and empties in the Gulf of Mexico. It starts from a lake in Minnesota. Iowa, Kentucky and Arkansas are some of the states through which the river passes. It runs for more than thousand miles and produces delta which becomes the bread basket of America. It is very broad river also. In certain areas it is even miles broad. Other rivers are also there. Other Missouri and Kentucky are its fellow rivers.

Mississippi delta produces wheat, soya beans and millet. Without this river there is no American prosperity. This river is so rich and wonderful that Mark Twain spins his stories around it. It is muddy and powerful. It touches Canadian Prairies also. It is the life blood of native American. But the white man stole it from the natives. There is wild life. It is major delta system in America.

Now the river is polluted by salt water infiltration. It is called mighty Mississippi. Shipping and farming are around this river. Spanish explorers found it. It has many tributaries. It is two hundred feet deep. It is also a base for bio-diversity. There is a beautiful city called St. Louis. It is located at the confluence of Mississippi and Missouri rivers. Egypt becomes a desert without Nile, America becomes a desert without Mississippi.

St. Lawrence River:

St. Lawrence story is both shocking and surprising. In the early stages of Christianity, he was a deacon in Rome. He was dedicated to the social service and religious duties. It was rumored that many diamonds and golden ornaments of Christian rituals were under his custody. It was said even holy grail and chalices were under his control. The governor had an eye on these precious materials like any other ruler of a country.

Hence, he demanded that the wealth of the church should be donated to him or to the state in the court. The deacon was in a state of shock and confusion. He asked the governor to allow him three days for collection of wealth of church to which governor granted permission willingly.

After three days he brought a small number of orphans, widows and disabled people to the court and said to the governor that those people were the wealth of the church. This infuriated the governor, who ordered that the defiant priest be put to death. He was burnt at the stake and became a martyr to the cause of Christianity. Indeed, blood of the martyrs is the seed of the church.

This river in Canada is called St. Lawrence River in the memory of the murdered deacon. There are five important lakes in North America. These lakes form basis of this river which flows into the Atlantic Ocean. It forms a clear border between Canada and the United States.

St. Lawrence River starts in Kingston. It runs for more than five hundred miles. Lake Ontario and St. Lawrence River are joint water ways. It goes through many cities like Ottawa and Montreal. The river freezes five times a year. The river of ice is St. Lawrence River. The ice breakers are used for ship movements. French man discovered this river. This is the second largest river in Canada.

It goes through Quebec. There are six locks in the river. Five locks are important. St. Lawrence River connects them and merges into the Atlantic Ocean. Lock is an elevator for the ship. Thousand islands are located in the course of this River. These islands are maintained as National parks. Quebec and Montreal are cities on the river banks. It is a messenger between the French speaking and English-speaking peoples of America North. St. Lawrence River is pride of Canada and beauty of Atlantic.

The famous Niagara Falls are only here. This river was discovered by Cartier in 1534. St. Lawrence River connects the five lakes with Atlantic Ocean through St. Lawrence seaway. It is a symbol of unity and friendship between The United States and Canada.

CHAPTER 5

AFRICAN RIVERS

Colonialism:

When I write about the rivers of Africa the Nile, the Cango and the Niger I am forced to make certain observations on the events of past history. History of the world is the history of the Europeans who began to dominate the other races and peoples after the discovery of sea routes and invention of gun-powder. Mariner's campus and printing also contributed to some extent in this terrible and somewhat ugly process.

Asia, Africa and America are wonderful continents with high mountains perennial rivers and green forests. The natives were living peacefully with their pastoral, agricultural and coastal backgrounds. They had their own social codes and religious systems. Indeed, all religions -Hinduism, Islam, Christianity and Buddhism originated in Asia. In Asia there were empires in India, China and Cambodia. They had peaceful rule and prosperous life.

South East Asia turned into a French and Dutch colony in the previous century. Europeans and white traders disturbed China through their opium wars. The religious dissenters moved to America in a ship across the Atlantic and disturbed the natives the Red Indians. Columbus and his followers killed many natives in South America with their guns and destroyed the remnants of Mayan and Aztec civilizations. The English and the French destroyed the aborigines and forest dwellers in Canada and created Quebec and Ottawa, Montreal and Toronto.

There was a peaceful continent, Australia. Captain Cook did not discover Australia. He reached Australia sailing across the waters and gave the news to Europeans. The natives were killed or driven to the forest and the continent became a penal settlement. All the criminals and anti-social elements who crowded English prisons were exported to these places and they established their societies in New Zealand and Australia. Moreover, one Magalleon of Iberian Peninsula sailed around the world and gave the idea of earth as a globe and removed the idea of the Atlantic as the end of the world. Such adventurers and sailors are described as the path finders and the explorers in European records and books. In reality they were all mad fellows who ventured into the sea and came upon places and islands by chance, which later on Europeans grabbed through gun-powder or trade. The coffers and treasuries of European Kings and Queens were filled with looted gold and silver, gems and diamonds of Africa and the new world. William Drake and Columbus were not explorers but buccaneers.

Thus, white man is a usurper and a colonizer. The natives were heroic and sometimes noble and generous also. The English traders bended in the Mogul court and sought permission for business. First, they wanted trade centers but later on they usurped India. Thus, white man has betrayed his hosts in every place. He has to pay back the natives his due today or tomorrow.

His only argument is that he has gone to these places only to civilize the natives and it is white man's burden. This missionary zeal is a sheer non-sense and Europeans went to exploit these continents and to enslave the natives. Colonization and slavery go together in white man's misadventures. Thus, colonization in the name of civilization had been going on for four hundred years in other four continents. Even great Shakespeare in his play Tempest makes fun of Caliban the native who accuses Prospero of usurping his island from him.

Africa was a beautiful continent. Cango Empire and Ethiopian palaces indicate the glory of African civilization. Alexandria was a centre of learning and trade. Silk Road gave wonderful things of Asia to Europe. Still white man looks down upon the Asians and Africans and claims superiority. Napoleon remarked that all beyond the Pyrenees were parts of Africa and not worthy of his conquest.

All their books and novels portrayed black man as an animal. To Conrad Africa is a heart of darkness. Europeans described it as a dark continent of cannibals diseases and high temperature. Slave trade in America and voyage of indentured labourers from Asia to the Caribbean's indicate the cruelty and exploitation and barbarity of white man. This is pure racism.

Superiority of white man is a myth and it is time that Africans and Asians must come together to break this myth. The book written by Edward Said, on Orientalism indicates white man's pride and his contemptuous term other to denote the blacks and browns.

In the post colonial period natives and aborigines must assert themselves and claim their ancestral lands and properties from ex-colonizers.

They should not fall prey to globalization which is another form of colonization. The computers, televisions and cell phones have removed the traditional tools of the natives. The multinationals and companies are trying to usurp the lands of the natives and destroy their original crafts, business and agriculture.

The American struggle for freedom and South American wars are not real movements of nationalism. In these historical events the power moved from European Governments to the white settlers. It is only transfer of power from Europeans to the diasporic. It is not freedom for the natives. The real freedom is the freedom of the natives and aborigines to own their lands to maintain their culture and to establish their rule. Partition of India, cruelties in Belgium Cango, the massacre in Greenwood in America, the genocide in Ravanda and Bengal famine in India are historic atrocities of colonialism.

The post colonialism is a process not of revenge or retribution but a process of rehabilitation. The myth of white man's supremacy must be broken. In the post colonial period African, Asian and American natives must come together and must claim their rights and lands. Robert frost is my favorite poet. But in his poem "Gift Outright" he says that land was ours before we became the citizens of the land. This is sheer colonial voice and I appeal for a new order in the world where blacks, browns, yellows and nomads must enjoy the resources equally and the myth of the white man's supremacy is broken.

Colonialism/anti-colonialism/post colonialism/neo-colonialism/globalization/ Racial Theory.

The above topics are not related to literary theories or economic principles. But they are related to dispersal of human species and their inter relationships and consequent economic exploitation.

There are five races in the world. They are Negroid, Mangoloid, Australoid, Caucasoid and Nomads. The Negroid refers to black people living in African continent. The Mangoloid refers to yellow people with flat noses who live in Himalayan regions, China and Japan. Australoid are the browns who lived in Lemuria or Kumarikandam and moved to other Asian countries after the deluge. The Caucasoid are the white people who lived in those mountains near Russia and spread across Europe. The nomads are the wanderers without any identity of region and Romanies, hunter gatherers and shepherds come under this group. I want to tell the truth of human races. These races lived peacefully in their regions. But the white people started colonizing other continents and conflicts arose amongst these races. Alexander and Caesar were the first colonizers of the world. Alexander from Greece destroyed Persian Empire and set fire to Persepolis the capital. Lord Russell describes him as the madman of Greece. Julius Caesar captured not only Cleopatra but also Alexandria and set fire to library, the first barbaric act. But the natives retaliated. Chandragupta of India drove away Greeks from India and Carthaginian General Hannibal retaliated and captured Rome. The modern man who retaliated is Robert Mugabe of Rhodesia in Africa.

Colonization went in full swing after the discovery of sea-routes and all the four continents came under the sway of Europeans. France,

Germany, Belgium, England and Netherlands were usurping regions in America, Australia, Asia and South America. The forerunners were the Portuguese and Spaniards. These Europeans had treaties and agreements in looting the regions of the world. Colonialism led to slavery, depletion of natural resources in colonies and dispersal of white man across the globe.

Anti colonialism is a process that started in 19th century. It was in full swing in 20th century. Many Asian and African counties became free but the white people created a bad atmosphere of tribal war, religious war and ethnic strife in the colonies. In the meantime, many associations like common wealth, UNO and UNESCO were started to maintain the supremacy and rule of the white man. For example, Pope in Vatican is selected by cardinals, who are whites but no brown or black has adorned that office so far. Olympics and tennis tournaments and cricket matches are nothing but revival of white man's supremacy.

Post colonialism is a process of stability of politics and economies in African and Asian countries. Post colonialism is the revival of native talents and creative power in Africa, Asia and the Caribbeans.

But neo-colonialism is a shrewd process of the Europeans. Colonialism is a day light robbery. But neo colonialism is a nocturnal theft. IMF, WTO, WHO and other bodies are working in favour of European countries. Among the colonies no unity prevails and Europeans sell their arms to these colonies at great profitable rates. When Asian countries like India and Pakistan go to war, more and more arms are sold to warring factions. Neo colonialism is followed by globalization in which multinationals and commercial groups flourish at the cost of native crafts, cottage industries and local markets.

In Africa and Asia, the pesticides and chemicals which destroy the crops are purchased from foreign companies. Small markets and fairs which sold local products and animals are pushed away by the Walmart, Amazons and Flipkart. The natives fall prey to computers, cell phones, televisions and you tubes and they lose money to the whites.

The continent of America, South America and Australia are still the white man's paradise. The natives and aborigines have no right to properties. 75% of lands and properties are in the hands of the whites and their organizations. In a Canadian school many children of the native population have been killed and buried. Pope has not expressed his pain over this though the Canadian Prime Minister expressed an apology.

The services of two white men, Albert Sweitzer in equatorial Africa and abolition of slavery by Abraham Lincoln in America can not wipe out the tears of the blacks and the blood of the browns. The fourth world of nomads, hunters, food gatherers and gypsies are living in the fringes of our prosperity and they have no citizenship, nationality and cultural autonomy of their own. Hence, I appeal to the world people to work for a new order in which all of us enjoy the fruits of our labour, resources of nature and benefits of millennium in equal manner without the supremacy of white race.

In South Africa Gandhi was kicked out of train by a white man because he was a brown. In America Martin Luther King was shot dead because he wanted his black children to sit at the same table with white children. Apartheid was practiced in South Africa for three decades before our own eyes. This is called racism.

But we should know that black musicians like Duke Ellington and sportsmen like Jesse Owens were better than white competitors. Tiger woods in golf and Monfils in Tennis are not white but blacks. Pyramids of Egypt, Temples of Asia and great wall of China are equal to the palaces in France, or any church in England. The colour of the skin is due to climate. It is not a god ordained system. All men and women are equal before gods of heaven and we should enjoy this human life without pride or prejudice. That is the will of gods and command of heaven. The American Tennis champions are not whites but blacks.

There are many things that contributed to the process of decolonization. In Africa or in Asia decolonization started from the European metropolitan cities. The educated Indians and Africans understood patriotism nationalism and sovereignty. They understood rights of human beings and consequently spread those ideas among their peoples. There was a literary movement of Negritude which questioned values and morals of white men. Then they had a pan African concept to unite the native tribes. The last aspect is the world wars which damaged the economy of European countries. Hitler and Mussolini, though they were dictators, disturbed the foundations of British and French empires. The European colonizers could not manage their colonies. Hence, they were forced to abandon their colonies. When Japanese attacked South

Asia, Lord Mountbatten could not resist. At one point of time the imperialists felt to save themselves was their main aim. Hence, they abandoned their colonies. **However, they created divisions and disputes among the natives so that they can control their colonies from a distance. For example, when English rulers left India, the estates of coffee and tea were in the hands of English men. Decolonization is the child of anti colonialism. But post colonialism is the consolidation of native political power and national sovereignty.**

Tulsa Massacre:

I will write a story about America which Americans themselves deny, but history will testify the authenticity of my story.

There was a place called Greenwood in America in Oklahoma State. It was beautiful and fine. It was called the paradise of blacks and the place had good doctors, lawyers and businessmen. It was called by Booker Washington the Black Wall Street. The whites were jealous of this place because it was the heaven for blacks whose fathers and grand fathers were once their slaves.

There was a report that a black boy misbehaved with a white girl in a lift of a huge building. The white racists were waiting for this ignition. Within ten hours the whole place was destroyed by the white racists. Fire was there, shooting was there and mob violence was there. The whole town was destroyed and mass graves covered the dead bodies of blacks more than three hundred in number.

No white American writer, author or historian has mentioned this so far and my favorite Robert Frost did not write anything about the tragedy of the blacks. **"His death of a hired man"** moved my heart but he does not write about the deaths of blacks who were successful through their hard work and intelligence. This is the treachery of the white man and barbarity of his society in America. This tragedy is called Tulsa Massacre of 1921.

Oklahoma was a black state and blacks had their identity in that state. Natives and blacks were land owners there. But President Jackson destroyed their dream and pushed white people and white guardians into the state and made racial segregation possible by Removal Act. Even Afro- Americans and Euro Americans who live in the U.S do not know anything about it. Lincoln enacted abolition, but the fact of existence of a secret group named Ku Klux Klan itself from the days of civil war is a blot on the white man's face.

Human Zoos:

There was one inhuman and barbaric custom among these colonizers and Europeans.

They took black people to European capitals and put them in zoos as if they were savages and direct descendants of monkeys and anthropes. They were trying to prove Darwin's theory.

The crimes of white men became adventures and even Karl Marx who thought himself a revolutionary spoke nothing about Africans and Asians. He spoke for the white laborers and never worried about slaves blacks and browns.

Did not Shakespeare in his **Othello** speak about racism and put African to death just because he loved and married a fair Venetian white lady?

The Bengal Famine:

Bengal is a fertile state in India and it is the rice pocket of India. But there was famine in 1943 and people died in Calcutta streets without food.

The British rulers diverted the grains to war fronts and allowed the people to die. The Prime Minister Churchill said 'I hate Indians and they breed like rabbits." This is his famous remark about Indians. The ungrateful Churchill did not know that he got his cigars from Dindigul in Tamilnadu of India and Indian soldiers died in the world wars to protect the empire inglorious. The British people looted the richest country in Asia and left it with wounds of partition and blood of Hindus and Muslims.

CHAPTER 6

AFRICAN RIVERS

The Nile River:

The Nile is the best river of Africa and its god's gift to the so-called dark continent. The river is the longest river in Africa and flows from the south to the north of Africa. It empties in the Mediterranean Sea and it is the beauty of the African continent. It passes through many countries, like Rwanda, Kenya, Sudan and Egypt. Egyptian civilization was based on the waters of Nile and Cleopatra was called fish of the river. It creates a fertile soil in Egypt and the people of Egypt understood the value of agriculture with the help of this river.

In flooded areas of the Nile the ancient Egyptians threw seeds and crops bloomed within a short period. The reeds Papyrus grew along the banks of the Nile and paper was produced from it. Great kings and queens lived in Egypt and they left their marks in tombs, pyramids, temples and manuscripts. Hence, we cannot underestimate the value of this river. Actually, Nile has two names – Blue Nile, White Nile. They converge in Khartoum the capital of Ethiopia. Both start in different places – but from great lakes. Many dams are built across the Nile and the floods are controlled to a considerable extent. Though Egypt has been converted to Islam, the people follow their old customs and traditional ways. Today the Nile River has become very touristic and river cruise is popular. The Nile comes from Greek word which means river. Tour of the Nile River includes a visit to the pyramid, a stay in the star hotel in Cairo and an adventurer's camel ride on the nearby sands to have a drink or to see a belly dance.

Water and fertile lands are basis of civilization and the most ancient civilization was founded on the banks of this Nile River. There is a big dam called Aswan across the Nile. The glory and the value of this river were forgotten when Suez Canal was constructed a century ago between the Red Sea and the Mediterranean Sea to reduce distance of ocean-going ships.

Indeed, the Nile has become a trouble spot now. The dams are built in Ethiopia for power generation and water flow is disturbed. Egypt is protesting, still the problem continues. We should remember great scholars and geographers lived in Alexandria and contributed to geography, astronomy and philosophy.

Both Sudan and America have interfered in this water dispute and have offered a moderate solution. Many of the Egyptian farmers complain of water scarcity and they are switching to other water – resistant crops. Many of the areas have become dry and parched and no crops grow there.

The agricultural basis of Egyptian economy is slowly drifting to trade along Suez Canal and tourism along the Nile River. There is also income from oil reserves. Cairo the capital and political centre of Arabian countries is losing its charm and power due to Arab Spring activities and the oil supremacy of fellow countries. The death of the Nile has fixed the death of Cairo's supremacy and so we understand that water is a vital power and river is a vital basis for glory.

Some explorers mention of one Yellow Nile which flows under terrain and mixes with Main River. But even this yellow Nile appears here and there like a pond and pool where African wild animals appear to quench their thirst. Africans who have lost their mineral resources through colonialism must realize the importance of their rivers and must become self-reliant through agriculture and tourism.

The Congo River:

Congo River gets its name from the kingdom of Congo located at its mouth. Congo passes through many African countries like Zaire and Tanzania. Wonderful animals like giraffe, lion, elephant and tiger live along the banks of this river. It is the highway in Africa.

This river is described by European explorers and writers in the negative terms. Joseph Conrad describes it as a heart of darkness. Another man describes it as a river of devils and diseases. These are unwelcome invectives the racial minded Europeans used to describe Africa and its regions to create a bad image. This river empties in the Atlantic Ocean and it is second largest river next to Nile in Africa.

Desert is not a desert so long as there is a water course. The people are not barbarians and hunters when they cultivate lands to produce cotton, sugarcane and sugar beat. Africa has been exploited by the whites and hence it is poor. Congo is a river of mystery and turbulence. Kinshasa the capital of Congo is on its banks.

Congo rain forest is equal to Amazon rain forest. It contains many species of birds, animals and aquatic creatures. These form the wild life and bio-diversity of Congo rain forest. There are many hydro electric projects along this river. It is a forgotten river because it is in Africa. There are many islands in the river. Longston falls are very famous. These European names like Longston falls and Victoria Lake are unpleasant to my spirit in an African atmosphere and landscape. Congo is a deep river. It is next to the Nile in its distance. It crosses the Equatorial line in two places. It gets rain in one part or the other. Hence it is a perennial river. Its rapids make boat rides impossible. Conrad wrote about this river and boat ride along the river.

It starts in Zambia. It is a navigable waterway. Its delta was exploited by Belgium. Many people of this region were massacred by Belgium king for extracting and exporting gold. These forests, animals and rapids are wonders of Equatorial Africa. Joseph Conrad in his heart of darkness writes about this river and regions. This river is the artery of the forest. More than hundred tribes live in this river region. This is representative of African continent.

The Niger River:

This is also an African river. It is a great river of Africa. Indeed, it is the river of the desert that gives life to many tribes of Africa. The river emerges from the country called Nigeria. The name of this river is derived from Nigeria.

Niger is a long river running across Africa. It empties into the Atlantic Ocean. Many cities like Timbuktu are located on the banks of this river. The river takes various names according to the country it passes through. The river is celebrated by these African tribes as a 'great river', 'grate water' and big river. The water flows across hot African desert, quenches the thirst of the people living in the dry regions. It is the life line of the poor Africans who need the river for transporting goods and travelers. Without Niger all tribes will perish. Agriculture is not possible in hot sandy soil.

The river avoids desert and merges in the Atlantic Ocean. Timbuktu is an important place in the Niger River region. Niger is out and out a desert river. It reflects African tribal life fully. Scottish explorer Maryoc Park explored this river from its mouth to the centre. His party and he were killed by the natives. Twenty native tribes live in the river region. But of late the river is spoiled by oil industry.

The Niger River is divided into three-Upper, Middle and Lower. The Upper Niger is full of rapids, so it is not navigable. But Middle Niger is navigable. There are many islands in this portion of the river. They separate these regions. In short Niger is a river of transport. It is not a river for agriculture. It runs for three thousand miles. Niger delta is rich in eco-system. It has many lakes and canals.

An American poet was very sympathetic to slaves in America. He writes in a poem called 'Slave's Dream' about the river Niger. The tortured slave before his death thinks of his free life in Africa and his walk along the Niger River. Long Fellow understood the pains of a slave. Americans were insensitive masters. Hence Niger is as good as other African Rivers, but it runs as a symbol of African wild life and tribal culture.

CHAPTER 7

AUSTRALIAN RIVERS

Murray River starts from Lake Hume and empties in Southern Ocean. The people are complaining about various uses of Murray water. Much water is taken out for industry and exports remarks an Australian citizen. Australia will be affected by this wrong exploitation of the river.

Murray and Darling Basin is food pocket of Australia and people point out the best flow from this basin. To speak precisely these rivers were free before the settlement of colonials in Australia. The white man has usurped these places from natives and at present natives have no role in the cultural, political and economic life of Australia.

Farmers and fishermen are disturbed by the river flow. Murray Darling basin extends to the South of Australia. The death of fish in these rivers are not due to draught. More water is for industry, it is loss for agriculture. Much grain comes from Murray Darling basin and it is the food packet of the country. Two million people live here. But rivers struggle to survive. Water management is a problem in Australia. There is no water flow in the lower stream. It limited water supply to farming. Murray Darling Basin suffers due to poor water management under Murray Darling Basin Authority.

Serpentine and Harvey are other rivers of some significance. But the fact is that Australian environmentalists are very conscious of the damage to the river system due to industry and commerce.

This Harvey River flows in the western part of Australia and it is noted for fishing. However, we feel sad that white settlers have usurped all natural resources and sent the natives to the woods and poverty.

All the rivers are to be protected and their water should be shared by the natives and the settlers equitably. There should be no discrimination in the use of water in this Australia. Land belongs to all. God has given water to be shared by us joyfully. This is my theory. This is a friend of dolphins. The Atlantic salmon is found in Serpentine River and angling is popular with the local people. This river flows through New found land.

The Murray – Darling basin is becoming very dry, for the water is used for environment and industry. For agriculture the water is not supplied in adequate manner. Hence the food bowl is becoming dry and empty. The basin experiences draught and flood alternately. Australians are more worried about the river water. Hence many fish varieties are dead due to pollution and the shallow river has become a playground for water sports. This is the tragedy of Murray Darling basin. The river is two thousand kilometers long. The cotton cultivation in the upper regions affects water flow in the down stream. The river suffers from greed and mismanagement.

CHAPTER 8
CONCLUSION & EPILOGUE

All rivers are equal. They may be big or small, they may be deep or shallow, they may be wild or gentle but all rivers are equal in the sense that they carry waters and give fertility to the soil and life to man. Water is common and universal source of sustenance of man. It can never be controlled and exploited by any country in the world. Hence water resources should be regulated by UN water commission if not totally controlled by it.

Rivers are living organisms; only an insensible man will think that water has no life. Water gives life to many species and plants; hence water is a super life giver. It should not be polluted and the course of the river should not be changed. Man made canals and dykes should not disturb the origins and courses of channels and rivers. Since fresh water is a natural right of any man, no government has the power to poison, spoil and pollute the fresh water either in the lakes or in the rivers.

The fish dies in polluted waters hence man loses this main food that comes from seas, rivers and lakes. When I say man has no right to pollute water, automatically he has no right to pollute the earth, air or space for his personal profits or adventurous activities. Man should not pollute universe with his useless and futile space-programs.

Throughout the world farmers fishermen and bank dwellers are affected by pollution of rivers. Hindus in India worship rivers but they throw corpses unto the Ganges River. We need a rational and sensible attitude towards rivers.

Industry, tannery and mining should not be allowed near the rivers. Rivers are tourist spots, trade routes and living places. In Netherland and Venice people live in boat houses on the canals. In Vietnam markets are on the boats. Hence rivers should be protected at all costs. Main culprits in destroying rivers and deltas are not private citizens but governments. By building dams and hydro electric stations on the rivers they do damage to rivers. Himalayan regions were spoiled by Tehri Dam. Seas oceans and lakes are common properties of man and hence governments have not total control over them. Cauvery basin was spoiled by ONGC in Tamilnadu State of India.

CHAPTER 9

LAKES AND POOLS

Lakes are origins of our rivers and streams. They are not spoiled even today. Except boating for pleasure and fishing for life man does not approach them with any commercial purposes. Pools are small water bodies where rain waters gather, they quench the thirst of our animals like goats, cows and camels.

In England Lake Districts are very popular for natural scenery and lakes. They are tourist destinations in England. Hence, we should accord equal significance to lakes and must take steps to protect their banks, fresh waters and fish cultures.

The city of Madras has eight lakes in its out skirts. But the government and companies have spoiled them by housing projects, hence the city suffers without drinking water every year.

On the other hand, a district in Tamil Nadu has many lakes and the agriculture is successful without canals and rivers in it. It is the gift of the lakes to the farmers. Lakes are our sustenance in every way. We should understand it.

In this treatise on rivers, we have to include the lakes which are small seas on the lands. We have seen five great lakes in North America. They are great reservoirs of fresh water. They are Lake Superior, Lake Ontario, Lake Eric Lake Michigan and Lake Huron.

These lakes are located on the borders between the U.S and Canada. Such lakes are located in other countries also. For example, lake Baikal in Siberia is famous throughout the world for its fresh water and fish culture.

In Kashmir in Srinagar there is Dal Lake a tourist attraction and a landmark in the city. In Jaipur and Udaipur there are many lakes and we can see all palaces and revolving hotels in these lakes.

In Australia and Africa there are lakes from where the rivers flow into the plains. The positive thing about these lakes is that most of them are fresh water lakes and only a few are salty and not tasty. **In India Lake Mana Sarovar is considered holy and it is dwelling place of Lord Shiva, the supreme deity of Hindus.**

CHAPTER 10

NATURE RETALIATES

Man is a cunning animal; he is a selfish creature. He might have created god or gods for his own comfort and convenience. But nature is different. It creates man and it protects him. It feeds him also. Hence poet Wordsworth remarks "Let Nature be your teacher. Look at his lines from his Up Up my friend.

> "One impulse from a vernal wood
> will teach you more of man;
> of moral evil and of good
> than all the sages can".

Thus, the English poet speaks of Nature and its grace. But Nature is ferocious and powerful. If man goes beyond the limit by destroying woods, killing birds and animals and polluting air and water, it retaliates. Nature will survive without man; indeed, all animals and birds will survive without man. But man cannot survive without nature, trees, birds and animals. They are independent, but man is dependant. Man does not realize this fact.

The forest fires, terrible cyclones, tsunamis, earth quakes, volcanoes and mountain slides are the symptoms of retaliation.

Today man is punished due to the retaliation of Nature. Today man is punished with many diseases like avian flue, swine flu, aids, Ebola, and Corona. Man is not a hill; he is a drop in the ocean. In the expanding universe, he is a worthless tiny tot.

Whitemen's Empires:

Another technique of the white man in colonization is naming the places, rivers and falls in his language. For example, we have Victoria Lake and Langston Falls in Africa. These people have nothing to do with Africa.

The harbour in the East coast becomes New York, the state becomes New Jersey and another state is New Hampshire. These places are in U.K. They have nothing to do with America. But white people give such European names to the colonies and try to obliterate the history, tradition and landscape of the natives. As their gifts for all these exploitations, they gave three things to the natives, first is the Bible, the second is wine and the third is syphilis. Europeans brought deadly viruses and bacteria such us smallpox, measles, typhoid and cholera for which natives had no vaccination. They died in course of time and native population was reduced to thousands.

Hence human zoos are a terrible crime of colonizers. It started in Belgium Congo. The Africans were taken to European markets and zoos to be displayed as exotic animals. In European cities many such expeditions were conducted to fill purses of traders. These Europeans and whites talk about human dignity and rights. In European cities many such exhibitions were conducted to delight the white crowds and to fill the purses of traders. The authors like Jean Paul Sartre and Albert Camus, who wrote about existentialism in Paris, never wrote a single word about the sufferings of blacks in Algeria a French colony. Simon De Beauvoir who wrote about rights of women, does not mention about women who suffered in sugarcane plantations of the Caribbeans.

These were the empires of Europeans, but they were empires of dirt, dust and degeneration.

An empire may be expansive but it should give protection and prosperity to its citizens. But these colonial empires sucked the blood of the natives of America, Africa and Asia. They exploited their labour in plantations. They abused women who worked in their estates. White man's rule is nothing but nasty and brutal.

Whatever shines in the streets of London, Paris, Madrid and Lisbon is not white man's work but the essence of Asians' and Africans' labour.

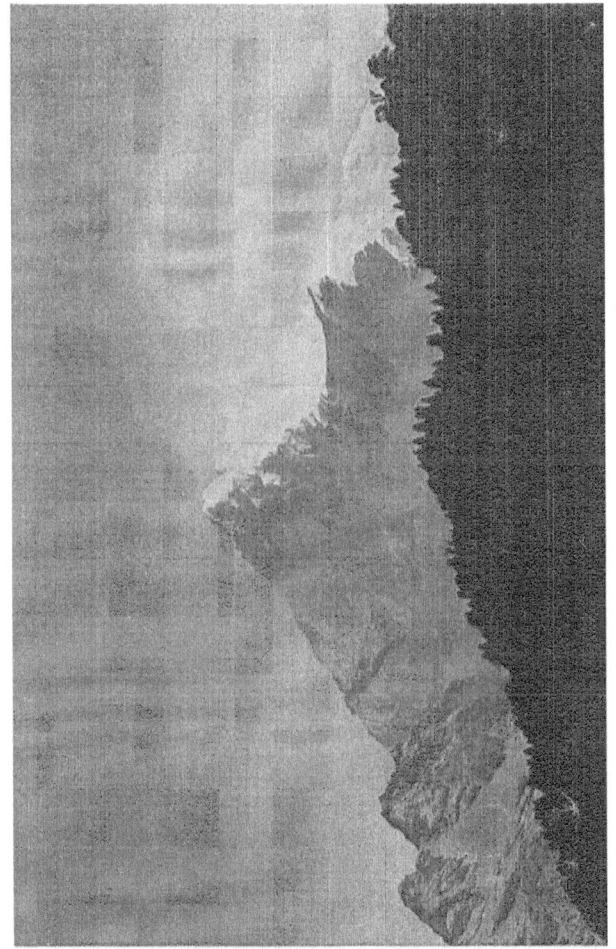

The Himalayas of Lord Shiva

The romantic Rhine river cruise

Capital Cairo on the Nile

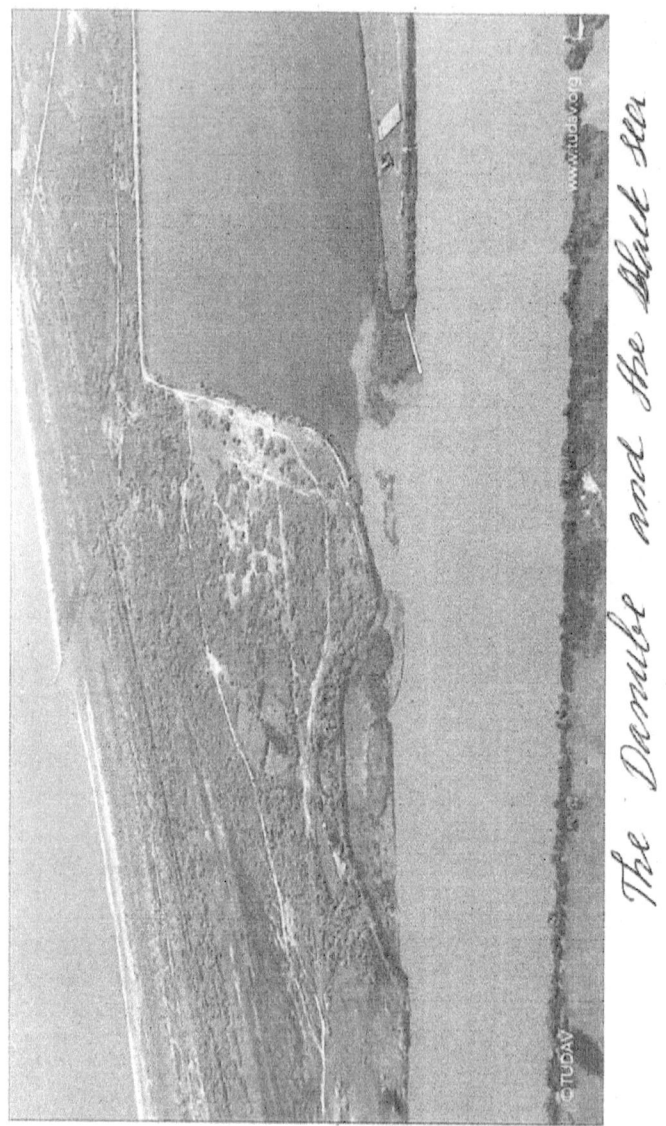

The Danube and the Black Sea

Haridwar on the Ganges

The boat market on the Mekong River

St. Louis city on the Mississipi River

The Volga Pioneer

The Rhone river valley

Budapest on the Danube

The Himalayan peaks

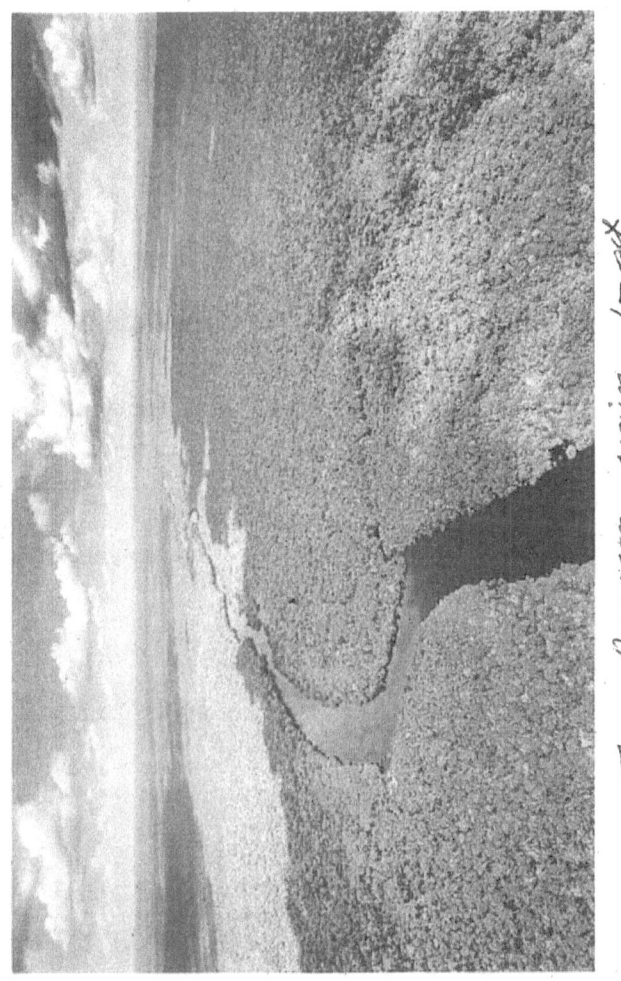

The Amazon rain forest

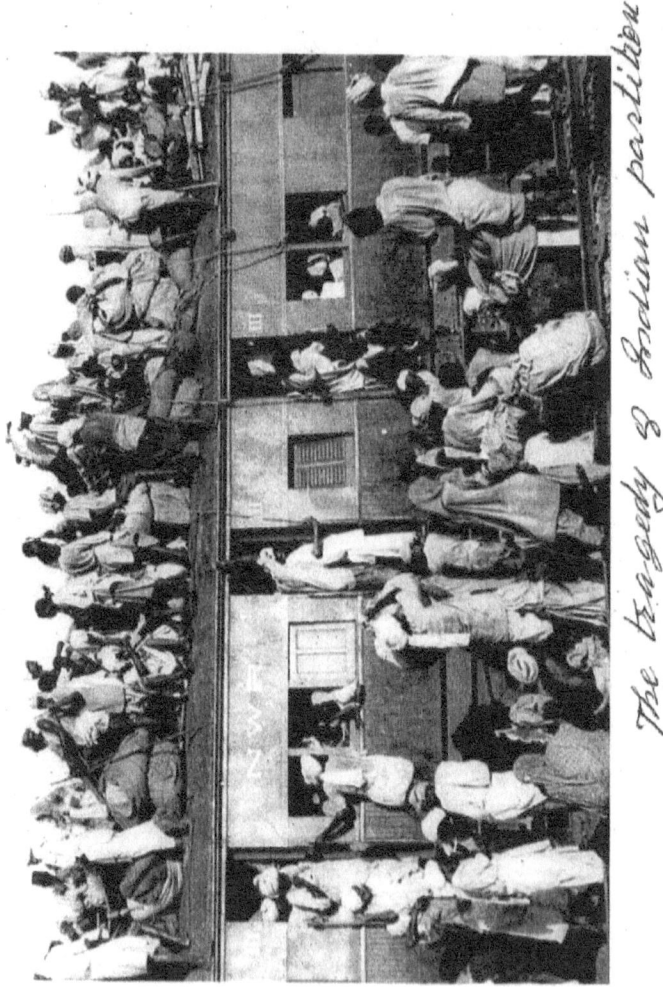

The tragedy of Indian partition

Tulsa Massacre in 1921
— Greenwood, America

The human zoo & colonial era

Murray Darling Basin

The Bengal famine of 1943

Congo River rain forest

The Romantic Rhine Cruise

Capital Cairo on the Nile

The Amazon Rain forest

Budapest on the Danube

Haridwar on the Ganges

The Boat market on the Mekong river